I0475533

Epitome
Of
Engineering
A Poetry
Collection

Math
Science
Innovation
Conflict
Conformance

Engineering
Problem
Solving
Essentials

By Ed Seymour

ISBN 978-1-300-34481-0

Preface:

Concepts embodied herein, hypothesis proposed and supported are deeply routed in decades of study. Topics included embody observations drawn on the study of team dynamics, cultural influences, technological innovation and legal practices that span from the 1960's to present day. An attempt will be made to delineate the differences between a digital immigrant and native.

The other principal point intended is related to key elements required to accomplish problem solving in our 24/7 internationally diverse world. We are inextricably interconnected by our continual efforts to leverage the latest innovation that has been observed and "do one better". In this vein, I will incorporate my observations as seen through the eyes of an active observer of trends which we have collectively seen from Punch Cards to Smart Phones and complex "ease of use" devices.

Ideas expressed within this poetry collection are opinions of mine alone and are not intended to express the views of any employer for which or with which, I have ever worked. My experiences in industry and academia have helped form this body of knowledge that I willingly share.

Throughout this book, I will emphasize the following Theorems and offer mathematical support thereof.

1. To error is human, to really make things interesting requires introduction of a computer.
2. If you think any product is really a commodity, reconsider.
3. Grass that appears greener may indeed be a function of the lens you use to see it
4. Saving for a rainy day really means don't waste what mother nature has to offer
5. Investment in the future must be longer term than financial "futures" suggest

To portray these topics in a fashion befitting will require multiple installments with this edition being the inaugural one. Suggestions and invitations to engage in discussions on all of the above are welcome and encouraged.

The last and final footnote on this topic, for now, is that to keep this to a containable size, I decided for the first edition to be a book of poetry, upon which I will expand into more, upon request.

Ed Seymour
PoemsOnTheSpot TM

Table of Contents

To error is human, to really make things interesting requires introduction of a computer.

Computers began with
1. Casio Databank, Punch Cards, Selectric Balls, Paper Tape, Magnetic Media
2. Room size to fits on your lap
3. Programming structures and login

If you think any product is really a commodity, reconsider.
1. New stuff becomes exciting
2. Once it catches on, it can be overwhelming
3. Humans look around to see what is being done by all competitors

Grass that appears greener may indeed be a function of the lens you use to see it
1. Examine with care what you already have
2. Do not fix what is already proven to work
3. Do not just sit around and do nothing

Saving for a rainy day really means don't waste what mother nature has to offer
1. Study what others do to conserve
2. Reuse, Reapply, Recycle
3. Keep it simple, where IT is a very flexible pronoun.

Investment in the future must be longer term than financial "futures" parlay
1. Real savings and real investment are not measured by 90 day increments
2. Real investment requires thoughtful dialog
3. Real is an important four letter word and subject to interpretation

Pyramids, Semiconductors and Engineering are 42
Clouds
Imagine

To error is human, to really make things interesting requires introduction of a computer.

We humans
Are much alike
Even
In the face of conflict

We
Admit
Openly
To making mistakes

When
We enroll a computer
In the task
Stuff can go MUCH FASTER

Computers Began With

Abacus
Pythagoras
Calculus
Succinctly stated, Math

Science
Added
Formulas
Predictions

People
Decided
On
Colors, sizes, shapes

Things
Have
and
Continue to evolve

Casio Databank, Punch Cards, Selectric Balls, Paper Tape, Magnetic Media

These
May have been called
Commodities
At the time of their use

Each one
Left an indelible mark
On
Our collective society

And
For that matter
Continues
To this day

To challenge
The norm
Question
Us

<u>Room size to fits on your lap</u>

Room size
Depends
On
The room

Laptop
May really be a question
of just how
Much you are willing to carry

Or
Perhaps
How fast and far
It needs to run

Before you
Prefer
To compute
In the Clouds

Programming structures and logic

At the end of the day
Logic
Is what makes computers
Tick

Based
We often think
On
Yes, or No

But then
There
Are
Exceptions

Rules
Meant
To hold
Until, broken

When
In fact
The outcome is more
Human

Humans
Gravitate
To
Scripts

Scripts
That
Govern
Plays

Scripts
That
Govern
Computers

Scripts
That
Govern
How we see the world

Scripts
Written
In
Perl

Scripts
Written
In
Basic

Scripts
Written
In
Java

Java
is
Sometimes
Confused with Beans

And
Often
Accompanied
By a cookie

And we all know
If you give
someone a mouse
And a cookie......

If you think any product is really a commodity, reconsider

Good ideas
Get used
Over
And Over

Really
Cleaver ones
Deserve
Patents

Patents
Seem to be
Good ideas
Well explained

So that
Others
Can
If they choose, follow

Or
As often
Seems to happen
Lead

New stuff becomes exciting

Who can resist
The urge
To see
The latest thing

That new gadget
That
You
May have only heard about

Being
Possible
Some day
Maybe

Yes
No
Maybe
As a new architecture

now
there is an idea
Requiring
an entirely different discussion

Once it catches on, it can be overwhelming

Imagine
Being responsible
For starting
A new trend

Once
You set that precedent
You need
To regain anonymity

Otherwise
You may need
To remember
Exactly how you did something

So that
Next time
A change
Is not seen, by others, as the next innovation

Humans look around to see what is being done by all competitors

We like to refer
To this
As
Comparison shopping

Perhaps
On occasion
Survival
Of the most fit

Once in a while
Checking out
The latest
Trend

Or simply
Trying on something
To see
If the shoe actually fits

Grass that appears greener may indeed be a function of the lens you use to see it

Light
Comes
From
Sources

Very often
Unseen
at
First

Sometimes
Ill
Understood
By many

But
If you simply
Observe
For a moment

Reflections
Refractions
Bending
Aside

The source
Appears
Even
If reflected

Examine with care what you already have

Care
Is that we all
Respect
And Deserve

Self
Care
First and
Foremost

For once
We see
How to meet
Our own needs

It becomes
Simple
To
Care for others

It is
However
Most difficult to see
What exists in our own space

Right there
In our own
Backyard
Every day, all day

Do not fix what is already proven to work

Define
Fix
Define
Work

No
Wait a minute
I have a better
Idea

Trust
Your
Gut
To tell you

Does it
Feel
Correct
In your heart and mind

Would your mother
Really agree
This
Is a good idea

Good
For you
And
Your family

If yes
Then
Do
It

If no
Think
About
Why Not

Do not just sit around and do nothing

Unless
Of course
That
Is what strikes your fancy

In the early
Morning
When
You are waking up

And
You hear
The birds
In conversation

Telling you
Of the day
They have already
Explored

Giving you a chance
To simply
Absorb
All that mother nature has to offer

Combined
With
Human
Nature

Saving for a rainy day really means don't waste what mother nature has to offer

History
Repeats
Itself
Often

Rain
Comes
To those
Who least want it, sometimes

One person's trash
Then
Becomes
Downstream treasures

Water
In most forms
Is conserved
Like entropy

Consider
Carefully
What to do
With this life giving resource

Study what others do to conserve

We
As a country of vast
Resources
Waste much of them

Not really
Because
We think
It is a good idea

Not really
Because
Of
Greed

Not really
to try
and
Pull a fast one

Not really
With
Mal-intention
Most often

More often
Just
Because
We fail to reflect once in a while

Don't
Take the time
To look, listen
Observe

After all
We
Are
Human

To be human
Requires
Mistake
To happen

Most of the time
Learning
Of lessons
Takes place

Most often
We
Admit to mistakes
Transgressions

Sometimes
We
Actually fail to see
What we should

Reuse, Reapply, Recycle

Think
Before
You
Discard

Think
About
Who
May better use

Something
You
Once
Saw as important

If
After thinking
You see
This is really trash

Then
Invite
Someone else
To prove it is not their treasure

By
Either
Donation
or

Placing it
In a container
Or
pile

Precisely how do you define IT

Is
It
Really
Information Technology

Is it
really
an
Impersonal Pronoun

Is it
Really
A person
With feelings

Is it
Something
Worthy of
More discussion

Yes
No
Maybe
Probably, Not

Keep it
simple
where IT
is a very flexible pronoun.

Investment in the future must be longer term than financial "futures" portray

To predict
The future
Is
An idea that holds out

Hope
For
Us
And is

An embodiment
Of
Our
Human Spirit

Some
Will insist
On doom
As the real deal

Some
Will
Imply
They know with certainty

What will
Happen
If
We fail to act

But
Those with better vision
See
What is here and now

Real savings and real investment are not measured by 90 day increments

Basic
Cash
Accounting Rules
Still apply

You want something
Save
For it
First

When you have the money
To buy it
Outright
Go ahead, buy it

You see something
With potential
Talk to a real
Bank

One
Willing to take
Calculated
Risk

Completely
Disclose
Intentions
Verb Subject Noun

Period
End
Of
Story

<u>Real investment requires thoughtful dialog</u>

Cite
Your
Intentions
Clearly without questions

Repeat
Yourself
One
and only one time

Ask
What was
Understood
As just said

Act
Accordingly
With
Decisiveness

<u>Real is an important four letter word and subject to interpretation</u>

Perception
a nd
Collective
Reality

Are
In fact
One
And the same

Facts
are
Words
We use to describe

The
Reflections
We
See

In plain
English
Real and
Reel, seem the same and different

Pyramid Semiconductor Engineering is 42

Look
At
A
Pyramid

Square
Base
Five
Faces

Add
Count
Of vertices
Total, 13

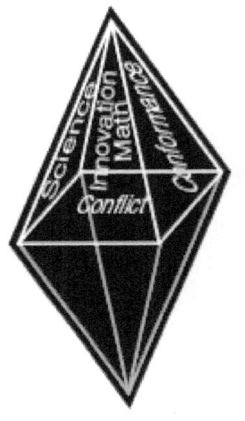

Add
Line
Segments
Total, 18

Render
In
3 dimensions, origami
Total 21 (black jack)

Reflection
around
Z axis
Diamond

Merged
Figure
42
Engineering

Clouds

These
Traditionally
Implied
Vapor, suspended

Far
Above
The
Earth

Holding
Moisture
Until
Release

Now
Clouds
Store
Bits, Bytes

Often
From
Collaborations
Joint Work

Coming
In the form
Of
Private

Shared
Local
Remote
Hosted

Using
Appliances
For
Engagement

Where
Transactions
Need to be
Secure

Security
Rests
In the
Provider

The
User
The
Entire Organization

Which
Implies
Doors
Open

Need to be
Monitored
Closely
Then, Security

Is
Much
Easier
To insure

Portable
Data
Ready
To be accumulated

Analyzed
Joined
Amalgamated
And Archived

All
In
A
Days work

At
The
Touch
Of a Screen

10/11/12

__Imagine__

The
Size
Of
Today's Switch

First
Pull out
A piece
Of your hair

It
Tends to be
One Hundred
Microns in thickness

Next
Ever wonder
How
Small a switch could become

In
Most cases
To first order
Lets imagine a transistor

And
For illustration
Lets take
Common size of 45 Nano Meter

Which
Requires some Mathematical
Manipulation
To see

If
You arranged them
Next to each other
How many could fit

100 microns
Is
100×10^{-6}
or 1×10^{-4}

45 nanometers
is
45×10^{-9}
or 4.5×10^{-8}

So
Simply divide
$(1 \times 10^{-4})/4.5 \times 10^{-8}$
And you can fit

2222
Transistors
There
On the width

Of
A Single
Human
Hair

And
When
You
Do

You
Cant
Really
See

What
Is going
On
Without

Bouncing
Electrons
Off
The shapes

And
Recording
Collisions
To sketch the landscape

This
Illustration
Shows
How far we have come

But does not
Even
Begin
To say where

We
Will
Venture
Next

As
Semiconductor
Implies
Not

A
Perfect
Switch
Not really, On, Off

More
Like
On a little
Versus, On a lot

Which
Explains
Ions
In Semiconductors

Ions
In
Batteries
Always want to keep Moving

Perhaps
We
Can
Find a special way, to contain them

If
We embrace
The
Idea that parasitics

Are
Really
Passive
Elements

Which
Can
Be
Used

To
Our
Advantage

Without
Much
Extra
Work

We
Then
See
How long term use

Of
Circuits
As
We now see them

Can
Be
Better
Much, Better

10/11/12

www.ingramcontent.com/pod-product-compliance
Lightning Source LLC
Chambersburg PA
CBHW021850170526
45157CB00006B/2388